THE UNIVERSE & GODS WITHIN BODY

SANDEEP BISHT

Copyright © Sandeep Bisht
All Rights Reserved.

ISBN 978-1-68487-562-7

This book has been published with all efforts taken to make the material error-free after the consent of the author. However, the author and the publisher do not assume and hereby disclaim any liability to any party for any loss, damage, or disruption caused by errors or omissions, whether such errors or omissions result from negligence, accident, or any other cause.

While every effort has been made to avoid any mistake or omission, this publication is being sold on the condition and understanding that neither the author nor the publishers or printers would be liable in any manner to any person by reason of any mistake or omission in this publication or for any action taken or omitted to be taken or advice rendered or accepted on the basis of this work. For any defect in printing or binding the publishers will be liable only to replace the defective copy by another copy of this work then available.

" THE UNIVERSE & GODS WITHIN BODY"

Contents

Prologue   vii

- THE UNIVERSE & GODS WITHIN BODY
- THE ULTIMATE GOAL OF LIFE
- MOKSHA & THE ULITMATE GOAL
- BHAKTI YOGA
- TYPES OF DEVOTEE (BHAKT) IN BHAKTI
- BENEFITS OF BHAKTI YOGA
- KARMA YOGA
- THE TYPES OF KARMA YOGA
- IMPORTANCE OF KARMA YOGA
- PRINCIPLES OF KARMA YOGA
- JNANA YOGA-THE PATH OF WISDOM AND KNOWLEDGE
- THE FOUR PILLARS OF JNANA YOGA
- LOVE ACCORDING TO SCRIPTURES
- RELATIONSHIP CONCEPT AND WHICH TYPE OF PARTNER WE CHOOSE
- NIRVANA
- POWERS BEYOND HUMAN BODY
- HUMAN BODY AND UNIVERSE ELEMENTS

Prologue

THE UNIVERSE & GODS WITHIN BODY

In the past decade there has been an explosion of interest in the 'body' as an analytical category in the social sciences and humanities, particularly within the context of cultural studies. Studies of the body have prolified, representing a range of disciplinary perspectives, including philosophy, anthropology, sociology, history, psychology, linguistics, literary theory, art history and feminist and gender studies. Despite the proliferation of scholarship on the body in the human sciences, until recently relatively few studies have focused on discourses of the body in religious traditions-on the ways in which the body has been represented, regulated, disciplined, ritualized, cultivated, purified, and transformed in different traditions. In recent years a number of scholars of religion have begun to reflect critically on the notion of embodiment and to examine discourses of the body in particular religious traditions. However, the body has yet to be adequately theorized from the methodological perspective of the history of religions.

According to my experience the body is represented in the Brahmanical tradition as a site of central significance that is the vehicle for the maintenance of the social, cosmic and divine orders. The body is the instrument of biological and sociocultural reproduction that is to be regulated through ritual and social duties, maintained in purity.

Most of humanity rightly believes that our Ultimate Goal is Happiness. They thinks Success is happiness. but that is not. Everybody wants to be happy, but unfortunately, not everybody is. We live with the sole purpose of doing things that can put a smile on our face. There are some who just exist, they drag through life. It seems like their life has no meaning, no purpose, and no goal. We all come into this world, live for about 5 to 10 decades and then depart. Very few really stop to nd out what is our true purpose. There are some who feel that they have discovered the secret that the goal of life is not just to be happy, but to make others happy too! Is this true?

Can we be happy all the time? While we all experience pleasure and joy, each one of us suffers misery and pain. We live with stress, fear, worry, anger and anxiety and this seems to be part of the life fabric that we are woven into. While we seek pleasure and shun pain, it seems that there is no way to escape from misery. Very rarely do we nd people who make this their Ultimate Goal, not just to enjoy bouts of happiness, but to achieve the state of Everlasting Peace and Eternal Joy without clouds of negativity spoiling their bright sky.

THE ULTIMATE GOAL OF LIFE

According to humanity the success and achievement is ultimate goal of life but that is not true. If anyone's thinks become a famous & financially strong is success.Those who thinks like that he or she is just a fool. The ultimate goal of life is free from that life cycle of human soul (i.e Moksha). Where we stuck this world. This world is just a jail or cage where we stucked need to free from this maya & material world. I also feels that there is nothing too much interseting in this world. When you gain this you will feel your life is balanced because it's my own experience.

Maya: "Gods powerful force that creates the cosmic illusion that the phenomenal world is real."

Under the influence of the three gunas, the soul is

- **Misled by matter**
- **Subsequently entangled**
- **Entrapped.**

This tendency is termed maya (illusion).

Under influence of maya's, the atman, (the soul) mistakenly identifies with the body. He accepts such thoughts as "I am white or beautiful and I am a man or woman," or "This is my house, my car, my country, and my religion." or " Unnecessary attachment with someone like He or She is my love (Boyfriend or girlfriend). These are just stuff that makes your soul weak and makes your soul impure. Love is not bad if both will love each other without lust. It is not necessary we need to love any man or woman. Basically we need to love all species & specially God. " Thus the illusioned soul identifies with the temporary body and everything connected to it, such as race, gender, family, nation, bank balance, and sectarian religion. Under this sense of false-ego (false-identity) the soul aspires to control and enjoy matter. However, in so doing he continuously serves lust, greed, and anger. In frustration he often redoubles his efforts and, compounding mistake upon mistake, only falls deeper into illusion.

In fact, Moksha is the Ultimate Goal of every human being, not just to live with peace but to be liberated from pain and suffering. Only a few are fortunate to go on a quest. They ask questions and investigate their doubts about life to ratify their beliefs. They are the ones who realize the Truth and achieve the Ultimate Goal of life.

Most of humanity suffers misery and pain. Not just the pain of the body which we all have to suffer, we also suffer the misery of the mind. As we age, our body tends to face all kinds of aches and pains. We also experience some diseases and ultimately the body dies. Nobody on earth can escape from this physical pain. Today, the world has advanced and we have medicines that can nullify physical pain. Not just ordinary painkillers, there are advanced opioids and drugs that can kill any kind of pain. If we can't reduce pain with medication, then we can use anesthesia to create an absence of physical pain. But what about mental distress? Who on earth is able to live without worry, stress, regret, anxiety, fear, revenge, guilt, and hate? We all experience such misery of the mind. While we are able to take painkillers to overcome physical suffering, how do we transcend mental pain? Then comes the agony of the ego. We are all subject to the ego making us miserable. We get angry, and upset over so many things. just because of the ego - the ego that demands and has expectations. All these together make us suffer the pain of the ego, the body, and the mind.

MOKSHA & THE ULITMATE GOAL

As I said before the ultimate goal of life is getting a Moksha and free from this material cage & maya (illusion).

Moksha: Moksha is the end of the death and rebirth cycle of soul. Permanently free from this material world. Moksha is the ultimate stage of salvation where the Atma, the divine body of Man, merges with Brahman, the ultimate reality. Brahman, which is beyond words or descriptions.

According to Bhagavad Gita there are three ways to achieve this Moksha:

- karma-marga
- jnana-marga
- bhakti-marga

Karma-marga: It is the path of action without selfish motives. Someone following this path lives in harmony with his/her ethical duty, or dharma, and may tend to be socially active. Doing good deeds is key to following this path. Karma marga is for those who seek salvation through day-to-day tasks while living an ordinary life working and raising a family.

Jnana-marga: Jnana-marga is a path of knowledge. Knowledge like Upanishads, Vedic scriptures, and the philosophic systems (as Sankhya, Vedanta, Yoga) and involving mental state. That also the path

In current time which humanity studying that's not a knowledge that just a information. That information is just a facts provided about something or someone. Knowledge is that which help's person to evolve themself. Maybe financially you will become strong because of that info. but you will not able to evolve because that info. .

Bhakti-marga: Bhakti-marga is the path of devotion, the method of attaining God through love and the loving recollection of God. Most religions emphasize this spiritual path because it is the most natural. As with other yogas, the goal of the bhakta, the devotee of God, is to attain God-realization--oneness with the Divine.

Bhakti-marga ways to do it:

- Chanting Gods name, mantras, etc to the Divine, either in a group or alone.
- Set up an altar with a favorite image or representation of the Divine and offer flowers, fruit, or incense; or do mental worship.
- Meditate on your chosen image of God.
- Choose a relationship with God that feels natural.

There are some types here in this bhakti yog.

- **Apara Bhakti (lower)** & **Para Bhakti (higher)**.
- **Ragatmika (bhakti withoutrituals)** & **Vidhi Bhakti (bhakti with rituals)**
- **Sakamya (devotion with desires)** & **Nishkamya Bhakti (selfless devotion)**
- **Vyabhicharini (devotion for both worldly attachments and God)** & **Avyabhicharini Bhakti (devotion only for God)**
- **Mukhya (primary)** & **Gauna (secondary) Bhakti**
- **Sattvic, Rajasic** & **Tamasic Bhakti**

In Bhagavad Gita chapter 9.26 ~ Whoever offers Me with devotion a leaf, a flower, a fruit or a little water – that, so offered devotedly by the pure-minded, I accept.

The term Bhakti originates from the Sanskrit word "Bhaj" which means love, attachment, faith, devotion, & prayer. Bhakti is very deep and intense emotion of love of the devotee for the Divine (Supreme energy). It's the purest, unselfish and most beautiful form of love where the devotee feels connected with God in his/her every breath.

A devotee in Bhakti loves God for the sake of love without any fear and selfish expectations. It's called "Parama Prem Rupa".

BHAKTI YOGA

Bhakti Yoga is little bit different from other yogs. It's also have some types which I already mentioned before.

Apara Bhakti & Para Bhakti

Apara Bhakti: Apara bhakti is the lower form of devotion where the devotee worships only his favorite God through all rituals and ceremonies. This idea of devotion is very narrow as the devotee disregards all other forms of God. This is rooted in desires and ego.

Para Bhakti: Para bhakti is the highest form of love for God which is pure and without selfish desires. A devotee loves God for the sake of love and always want to serve the Lord without any expectations. Para bhakti recognizes the transcendental nature of God.

Ragatmika bhakti and Vidhi Bhakti

Ragatmika Bhakti: Ragatmika bhakti is a free flow of love for God without observance of any rituals and ceremonies.

Vidhi Bhakti: Vidhi bhakti is devotion which follows rules, rituals, and ceremonies.

Sakamya and Nishkamya Bhakti

Sakamya Bhakti: Sakamya bhakti involves worshipping God to fulfill personal desires of health, wealth, and other material gains. God grants all wishes if the devotion is intense and prayers are done with genuine heart. However, the devotee never gets ultimate satisfaction and liberation through Sakamya bhakti because there is always a selfish desire and the love for God is not unconditional.

Nishkamya Bhakti: Nishkamya bhakti is the highest form of bhakti one should aspire for. Here, the devotees love God with the purest heart without any desires and feel his presence at all times. The divine grace bestows upon the devotee and he gets all the divine gifts (wisdom, power etc) from God without asking anything.

Vyabhicharini and Avyabhicharini Bhakti

Vyabhicharini Bhakti: In Vyabhicharini Bhakti, devotee's love is divided among family, material possessions and God.

Avyabhicharini Bhakti: In Avyabhicharini Bhakti, the devotee loves only God.

Mukhya and Gauna Bhakti

Mukhya Bhakti: In Mukhya (primary) bhakti, God is the primary aspect of devotee's life. And devotee's love for God is pure and spontaneous.

Guana Bhakti: Guana (secondary) Bhakti is devotion as a secondary aspect of devotee's life and he loves God according to his attributes or Gunas.

Sattvic, Rajasic and Tamasic Bhakti

Sattvic Bhakti: Sattvic Bhakti involves devotion to please God. The seeker only aspires for God and has no desires for materialistic concerns.

Rajasic Bhakti: Rajasic Bhakti involves devotion to seek material riches.

Tamasic Bhakti: Tamasic Bhakti involves devotion to achieve success through unfair means.

For example, a thief praying to God for success in a robbery.

A devotee should aspire to progress from lower forms to higher forms of Bhakti (Para Bhakti) where he loves God without any desires.

Based on different types of Bhakti, there are also different types of devotee (Bhakt).

How Bhakti Develops for God

First, a devotee feels absolute faith in his God. Then, the devotee feels intense admiration and attraction for the Creator (God). He gets drawn more and more to know, worship, pray and love God. Gradually, all his mundane worldly desires drop and he feels contentment and single-mindedness. Once that level of devotion is reached, the devotee lives every moment of his life in remembrance of God.

In the highest level of Bhakti, all attachments of the devotee with worldly objects fade and he feels attracted only towards God. This leads to a state where devotee feels one with God.

There are 9 stages in which devotion (Bhakti) develops in Bhakti yoga in a person.

9 Stage Process for Developing Devotion to God

- **Shraddha–** Faith
- **Satsanga–** Associate with devotees and spiritually advanced people
- **Bhajana-kriya –** Perform and participate in devotional service
- **Anartha-nivrttih–** Become free from unwanted material desires

- **Nishta** – Stability in devotional practice
- **Ruchi** – Develop a taste for bhakti
- **Asakti** – Develop attachment to God
- **Bhava** – Feels various emotions of love for God
- **Prema** – Pure love for Krishna.

TYPES OF DEVOTEE (BHAKT) IN BHAKTI

In Bhagavad Gita chapter 7 ~ Four types of men begin to render devotional service unto me (God) — the distressed, the inquisitive, the seeker of material wealth, and the one who has already realized knowledge of the Absolute

The types of devotee is:

A river gives water to everyone, however, how much water you get depends upon your vessel. Similarly, God has equal love for all beings. However, based upon unique temperaments of all individuals, each devotee has a different experience of the divine love.

People approach God for different reasons. There are four types of devotees:

- Artha (the distressed)
- Artharthi (the seeker of material things like wealth, health, house, love, etc.)
- Jijnasu (the inquisitive)
- Jnani (the self-realized)

Artha Devotee: Artha devotees remember God to alleviate suffering. When the problems get solved, these devotees again go back to their state of doubt about the existence of God.

Artharthi Devotee: Artharthi devotees want material things like wealth, family, fame from God. Their devotion to God is for the fulfillment of desires and wishes.

Jijnasu Devotee: A jijnasu is curious to know about God through inquiry and study of scriptures.

Jnani Devotee: Jnani bhaktas are self-realized devotees who know the ultimate truth of life i.e. God (Brahman). They are in touch with their divine nature and totally immersed in the love of God.

Emotions of Bhakti

When it comes to practice Bhakti yoga, the first thing pops up in the mind the is our emotions. In devotion, a person has the following 5 kinds of emotions which plays a very important role in the practice of bhakti-yoga.

- **Shanta**– Peaceful emotion
- **Dasya** – Servant devotee
- **Sakhya**– Friend Attitude
- **Vatsalya**– Motherly emotion
- **Madhurya**– Beloved emotion

Shanta Bhava: The devotee is calm, poised and peaceful. He does not exhibit many emotions, however, his heart is full of intense devotion. He silently and peacefully loves God with his heart full of love and joy.

Example: Bhishma. All renunciants have Shanta Bhava.

Dasya Bhakta (Devotee): When a devotee aspires to serve God whole-heartedly with a servant attitude, this is known as Dasya Bhav.

Example: Sri Hanuman used to serve Lord Rama whole-heartedly like a faithful servant. He found joy and bliss in the service of his

Master In the holy city of Ayodhya, the vast majority of people worship God with dasya bhava. Their names are like Ram Das, Siyaram Das.

Sakhya Bhava: In Sakhya Bhava, the devotee loves God like a friend.
This bhava is difficult to experience as devotee and God are on equal terms as friends. This bhava demands purity, understanding, openness, and courage to experience a relationship of a deeply intimate friendship with God. This Bhava can be attained only by people who are very mature and developed in Bhakti.

Example: Relationship between Arjuna and Lord Krishna.

Vatsalya Bhava: The devotee loves God as his little child. The devotee loses all fears and selfish desires in this Bhava as a mother cannot be afraid of her loving child. Nor can she expect anything from a small son.

Example: Yashoda's love for her son, little Krishna.

Madhurya Bhava: The devotee shares a relationship of the lover and the beloved with God. This is the highest form of Bhakti. The devotee and God feel one with each other while still being separate. Madhurya Bhav is totally different from earthly love as the former is selfless love for Divine while the latter is a selfish based on ego needs.

Example: Lord Gauranga, Jayadeva, Mira, Andal and many more.

BENEFITS OF BHAKTI YOGA

When we follow the path of karma yoga, it automatically develops caring nature within us & with caring nature, one can enjoy the fruits of Bhakti yoga.

Less dependence on the external world

A devotee is less dependent upon the external world –relationships, situations, fame, money, etc. for love and happiness. He has found an ocean of eternal peace and love through his unselfish devotion to God.

Removes all doubts and fears

Bhakti removes all fears of the devotees, even the fear of death as a true devotee only aspires for the love of God and wants nothing else.

Cultivates compassion and love for everyone

A devotee has equal love for all beings as he sees the Divine in all forms of creation. All negative emotions arising out of Dvesha (strong dislike) evaporate. He cannot bear negativity for any living being even a poisonous snake. He will keep an appropriate distance from the source of pain, however mentally he will maintain the love for one and all.

Provides emotional and mental stability

A devotee remains neutral through the thick and thin of life and sees the difficult situations as a part of the process of spiritual growth. A true devotee sees all events of life as drama and result of Prarabdha Karma (Past life Karma experienced through present life). Hence, when things do not go as per plans and desires, a devotee never loses his emotional stability. There are no signs of anger and painful emotional outburst in a true devotee. He respects the opinions of others as different points of view.

Dissolves the ego

A true devotee has no attachment and sense of ownership for people, possessions not even his body. A devotee's purpose of life to serve God and he feels an instrument in the hands of the Divine. He lives in a state of absolute surrender to God.
A bhakti yogi performs all worldly karma without any desires as he has found the eternal bliss and joy of the divine love in his heart. He is content within himself and hence, is not distracted by the power of worldly Maya.

Provide the ability to discriminate

A devotee develops an ability to discriminate between temporary happiness (Maya) and eternal joy (Divine love). A devotee observes and passes through all facets of the external world (pleasure, pain, beauty, novelty etc) without attachment. He knows the ever-changing reality of worldly happiness. Hence, a devotee knows that the knowledge of the ultimate reality, Brahman is the only security.

Purifies the heart

A devotee never feels envious to see anyone more blessed than him as he sees God in every being. He wishes good for one and all.

Improves confidence

A devotee is truly confident in him as he is and he never tries to please anyone. He is equanimous in all situations and gets along with all kinds of people and situations quite effortlessly.

KARMA YOGA

Karma yoga is not just that simple which humanity thinks. People thinks just doing there work is Karma. They really fools who thinks like that. The Karma yoga teaches us how to live happily while being in the hustle of daily life. Serving attitude towards the 'karma (work)' opens the gate of spiritual liberation (moksha) for a yogi. Karma consists of action we perform consciously or unconsciously & result of that action.

Karma (action) is not only the physical work but the process of mental thinking also. When Yoga is added to karma, it becomes a practice of union with one's true self through 'action'.

"Hence, every action which brings our awareness inwards to knowing the true self is the part of karma yoga."

"Yogah Karmasu Kausalam" ~ Bhagavad Gita 2.50

Yoga is an art of getting perfection (kausalam) in every work (Karmasu) of life. This perfection comes in karma with the regular practice of devoting karma to others. Hence, Perfection in karma is considered yoga also.

The path of karma yoga emphasizes on doing 'selfless work' according to dharma (moral duty), not on the consequences of results. A karma yogi treats 'work (karma)' like prayer where there is no outcome desire expect from karma.

THE TYPES OF KARMA YOGA

Yoga is not karma, but it's the practice to go beyond the karma. Not every karma is considered yoga. Hence, Types of Karma Yoga shows how many ways karma affects us.

Karma based on these 2 category.

- Intention
- Timeline

Karma based on the intention

Sakama Karma: Every single thought or physical action that creates 'mine' or 'your' sense in the person's mind is Sakama karma. This karma reveals the selfish nature of a person.

Yoga is not mean for Sakama karma because it keeps us bound in the bondage of karma, while yoga (Karma yoga) freed us from the bondage of karma.

Sakama Karma creates egoism, hatred, jealousy in a person's heart consciously or unconsciously.

Nishkama Karma: Nishkama means Selfless action. The good thing about selfless action is that it breaks the bondage of karma & let ourselves free from the cycle of death & birth.

When intention behind karma is not the main aim, a person actually imbibing the path of Nishkama Karma.

Forgiveness, helping, loving thoughts and compassion behavior of humanity is the example of Nishkama or selfless action. The practice of Nishkama karma leads seeker towards renunciation, which further purifies the Chitta (Mind/Thoughts/Emotions).

Karma based on the timeline

An action we had performed in the past, performing at the present moment & the result of the action we will get, is another base to categorize karma yoga. On this base of that.

There are three types of Karma Yoga.

- **Sanchita Karma**
- **Prarabdha Karma**
- **Agami Karma**

Sanchita Karma: Sanchita is the karma we had performed in the past. The literal meaning of Sanchita is 'Accumulation'. Hence, Sanchita is the set of accumulated actions of the past.

Every person has to go from some set of Karma in their life. As we live in the present, a conclusion (or rest) of this karma starts accumulated as Samsara 1.

sanchita karma is that glimpse of whose can be seen in the present 'Character' of person.

It is the **Law of Karma**– What we do in present accumulate in the past (as Sanchita karma) & appears in our future.

Prarabdha karma: Prarabdha is that part of karma (performed karma) which is responsible for the present condition of a person.

If you are experiencing something good this moment, it's undoubtedly because of the past karma of yours. Prarabdha karma is only can experiences whether it's good or bad, not changed. It's a debt of our past karma (Sanchita).

Agami Karma: Agami means forthcoming. This karma is the result of prarabdha karma. It can be modified according to the present of our working.

Out of these 3 Karma, Sanchita and Agami karma is not in our responsibility right now.

Krishna tells, Prarabdha Karma is the only responsibility of a person who decides everything.

Karma yoga demands you to work on Prarabdha karma only (Present action), as it's only our moral duty to focus on present condition or work, not past (Sanchita) or future (Agami) karmas.

IMPORTANCE OF KARMA YOGA

Source of Activeness– the practice of karma yoga brings activity in every aspect of life. Activeness is the key to self-development as it makes us realize the suffering or pleasure of life.

Teaches to Be Even-Minded– When awareness is detached from the karma's result, then yoga teaches us how to be evenminded in any condition. A balanced mind is the sign of a calm person that helps one during tough times to come out of sorrow. Hence, detached Karma is the quality factor of the mind's calmness & positive psychology.

Helping Ourselves by Helping Others – A right Karma is helping others. But, in the process of helping others, we help ourselves ultimately because it fulfills our heart with joy & perfection.

One Pointed Consciousness– We say something, do something, show something, and something else is going on inside us. It makes our consciousness shattered. Regular practice of karma done with rightful intention removes all these demerits.

Importance of karma yoga in Business Ethics

Achievements of Business Goals – To Achieve the ultimate aims of any business administration one should practice Karma Yoga. This is not only for entrepreneurs but also for the employees working under them.

Achievements of Personal Goals – According to Karma Yoga, while performing any task forget about your personal benefits. Put all your determination on organizational goals. In this way, you can achieve administrative and personal goals as well.

Appreciation – While doing any task just forget about your personal objectives and try to recall the business objectives. Do your duty with full determination and all your efforts. When you give your 100% then u will surely be appreciated.

Increases Catching Power– Through Karma Yoga, we gain the power to think positive and if we are thinking positively, then we will not take our duties as a burden and we will learn new things faster.

Active Involvement– This form of Yoga helps to direct the mind. A businessman is always a motivator or a leader who directs their employees in a positive way for the achievement of the ultimate goals. This increases the involvement of the employees in any task.

PRINCIPLES OF KARMA YOGA

It's very easy to emulate the practice of karma yoga in your daily life with these:

Start serving yourself

Serving to the self is the beginning step to start the journey of karma yoga. Whether it's work of a household or corporate life, specify your job & don't let others do it for you. This practice will keep you active & build a foundation of serving others too.

Be a genuine person

A tendency of faking to the karma makes a wall of duality in personal life. Most of the time peoples show something and something else is going on inside them. Presenting the true nature in front of others helps you to be conscious of the present moment.

Love your work

We know the Law of Karma, i.e. what we do in the present, is reflected in the future. Further, Prarabdha Karma is only what is in our right. Loving the work prepares a firm foundation for tomorrow (agami karma).

Practice to be compassionate

A few words of compassion can positively affect many critical conditions. Compassion came as an integral part of karma yoga when the seeker absorbed into karma. Practice compassion to mankind, animal and nature, and you will observe good vibes around you in every condition.

Forgo the result

To practice karma yoga, one of the keys is letting go of the outcome of karma. Krishna tells Arjuna in Bhagavad Gita 'Do your work, it's your duty & leave the result of work on me (God)'.

Respect the consequences

Some people don't let accept the result in a certain situation of life. Try to calm the mind in the critical situation of life and become a silent observer of karma. Respecting the consequences of a result and again doing karma with the same enthusiasm is like respecting God's offerings.

JNANA YOGA-THE PATH OF WISDOM AND KNOWLEDGE

Jnana yoga, a yogi seeks to achieve the ultimate goal of yoga by acquiring knowledge through scriptures and experiences of real life.

It is considered to be the most difficult path to attain self-realization because Jnana yogi requires to have an intense spiritual practice and discipline. Through meditation, self-inquiry, and contemplation, the yogi can attain wisdom about the true reality of self and be liberated from the Maya (illusions).

Jnana Yoga is also called the Yoga of Intellect as it is through knowledge of scriptures and self-study, one can unify the Atman(inner self) to the Brahman(ultimate reality). Through techniques of self-inquiry, conscious illumination, and reflection, defined in the Four Pillars of Knowledge, it requires the mind to move beyond intellect and seek the absolute truth.

Philosophical roots

The Jnana Yoga finds its root in the Bhagavad Gita and Upanishads. The Gita defines Jnana Yoga as the path to self-realization and the Upanishads underline the realization of the oneness of the self with God. Additionally, Bhagavad Gita highlights Jnana Yoga as a non-dualistic tradition of the Advaita Vedanta philosophy.

(The word Advaita means non-dual and Vedanta means Vedic knowledge.)

As per this philosophy, the knowledge acquired through the 4 pillars will bring the realization that the inner self is not separated from the ultimate reality. This liberation of the illusion of duality will bring an end to all your sufferings.

Moreover, the concept of Jnana Yoga was propagated by the ancient Indian philosopher Adi Shankaracharya, who consolidated the Advaita Vedanta around 700 BCE. According to his understanding, the yogis needed to completely renounce the world to be liberated from Maya to achieve self-realization.

The concept of Brahman in Jnana Yoga

Brahman, as outlined in Jnana Yoga, is absolute, the ultimate reality. The concept of space, time, and causation is unchanging and it has no beginning and end. The Brahman is infinite and it is something our normal mind cannot comprehend.

It is the highest universal principle that is omnipresent, which means it is present in EVERYTHING. As per Advaita Vedanta, Brahman is the ultimate truth that binds everything together in the universe. This concept teaches us that all people are spiritually one irrespective of caste, ethnicity, race, or nationality.

The Maya or illusion, which hides within the ego of mind and body, is the root cause of all sufferings and it is what separates us from knowing the Brahman. The practice of Jnana Yoga will help in making a connection to the Brahman by eradicating ego and liberation from desires and objects. The aspirant will be able to move beyond the illusion and experience a shift in their viewpoints and awareness.

THE FOUR PILLARS OF JNANA YOGA

Before embarking on the journey to self-realization, you must follow the Sadhana Chatushtaya or Four Pillars of Knowledge. These steps should be practised in a sequence as they are built upon each other. These pillars will provide spiritual insight and understanding and also aid in reducing the suffering and dissatisfaction in life.

1. **Viveka** – This Sanskrit word means discrimination and discernment. One should continuously and deliberately make an intellectual effort to distinguish between the Self and not-Self, the real and the unreal. Constant association with saints and continuous study of Vedic literature can help you develop Viveka to the maximum degree.

2. **Vairagya**– It means detachment and dispassion. A Jnana yogi should be non-attached to the pleasures of the world and heaven. However, it doesn't mean that you should leave everything and live a life of solitude in the Himalayas. You should be detached mentally from worldly possessions while carrying their duties and responsibilities. A long-lasting Vairagya can be achieved with a successful Viveka.

3. **Shatsampat**– They are the six virtues of mental practice to balance the mind and invoke discipline.

The six mental practices are:

- **Sama** – serenity or tranquility of mind brought by diminishing any desires.

- **Dama**– restrain of control of the senses to be used as instruments of mind.
- **Uparati**– a natural renunciation or withdrawal from all activities except one's duty or Dharma, that will be achieved after Viveka, Vairagya, Sama and Dama.
- **Titiksha** – the forbearance or endurance of extreme opposite states such as hot and cold, pleasure and pain, etc.
- **Shraddha** – having faith and trust in the guru's teaching, scriptures and self, through reasoning, experience and evidence.
- **Samadhana**– the focus and concentration of the mind on the Brahman or Self. The aspirant will enjoy a greater peace of mind and inner strength when practicing the above 5 virtues.

4. **Mumukshutva**– It means longing or yearning. The intense desire for liberation from the wheel of birth and death, sufferings, sorrows, delusion, old age, and diseases. If the aspirant has successfully practised the Viveka, Vairagya, Shatsampat, Mumukshutva will easily come to them. When one achieves the purity of heart and mind along with discipline, the yearning for liberation dawns by itself.

How to Practice Jnana Yoga

It is said that people with pure hearts, open and rational minds, and sharp intellect can take the jnana yoga journey. To practice this path of yoga, a person should first practice Karma Yoga and Bhakti Yoga to prepare the heart and mind to receive the Knowledge. It is recommended to practice Jnana Yoga under the guidance of a highly expert and qualified guru to keep an accurate track of your progress.

Once you have successfully completed practising the Four Pillars of Knowledge, you are ready to practice Jnana Yoga in its essence.

These practices have also been in the Upanishads, which are:

Sravana – Hearing or experiencing the Vedic knowledge and Upanishads literature through a guru. Here the guru will impart all his knowledge and the philosophy of non-dualism to their disciples through various analogies and stories. The student also studies the Upanishads on their end to assimilate the knowledge to understand the concepts of Atman and Brahman.

Manana– After the disciples have attained all the knowledge, they should now contemplate on it. The students should reflect and observe the teachings received from the guru and derive conclusions from them. They should think about the concept of non-duality and understand their subtleties.

Nididhyasana– The last practice is meditation. Here the student performs constant and profound meditation on the Brahman, which ultimately leads to the expansion of Truth. Through meditation and reflection on the primary mantras of the Upanishads, the aspirant can pursue the union of thought and action.

Stages of Jnana Yoga

The stages through which a Jnana yogi will progress have been described in 7 stages by Swami Sivananda known as Jnana Bhumikas. It is a roadmap through which the yogi can gauge their progress and follow the path of self-realization.

The 7 stages of Jnana yoga are:

Subheccha (good desire) – The first stage will be achieved after intense Sravana and performing righteous action without expecting any return. Through this, the mind will be cleansed of any discrimination and non-attraction to sensual objects will prevail. This stage can be said as the foundation for the next 2 stages.

Vicharana (philosophical enquiry) – It is the stage of constant questioning, reflection and contemplation on the principles of non-dualism.

Tanumanasi (subtlety of mind) – This stage is also called the Asanga Bhavana as here the mind is free of any attractions. It is assumed that the aspirant has understood all the knowledge imparted by their guru and their mind has become thin like a thread (Tanu meaning thread). If a yogi dies at this stage, they will stay a long time in the heavens and will be reborn as a Jnani.

Sattvapatti (attainment of light) – Here the world will appear like a dream and the yogi will look at the things of the universe with equality.

Asamsakti (inner detachment) – Any leftover desire is diminished in entirety in this stage. There is no difference between the waking and sleeping stage and yogi experiences of Ananda Svaroop (the Eternal Bliss of Brahman).

Padartha Bhavana (spiritual freedom) – At this stage, the yogi will start understanding the Truth and the Brahman (ultimate reality).

Turiyatita (supreme freedom) – This is the final stage where the aspirant has attained Moksha (liberation, enlightenment). The yogi has achieved the state of superconsciousness and Videhamukti (liberation without the body).

LOVE ACCORDING TO SCRIPTURES

(Bhagavad Gita 2.13) Krishna starts His message of love by enlightening Arjuna: "We are all souls, spiritual beings, entitled to rejoice in eternal love with the supremely lovable and loving God, Krishna." When our loving nature is contaminated by selfishness, we start loving things more than persons, especially the Supreme Person.

The Sanskrit words for love is **"sneha"**. Other terms such as Priya refers to innocent love, Prema refers to spiritual love. the Vedic people developed a human relation basically related to love. Love had an important role in their society. Love was as natural as it could be. It talks about body, soul, relations, harmony and even jealousy and domination.

There are very interesting origins of love in Hinduism (Hindu Dharma). One origin has it that there used to live a super-being known as 'Purusha', who had no desire, craving, fear, or any impulse to do anything as the created universe was perfect and complete.

Then Lord Brahma, the creator, took things upon his hands and split Purusha in two. Then, the Sky separated from the Earth, the darkness separated from light, the life separated from death, and finally, the male separated from the female. Each of the halves then went on a quench to satisfy itself by reuniting with its other half.

There also some stages:

<u>Kama – Sensory Craving (that is not love according to me)</u>

The desire to merge initiates from physical attraction. The Kama means "craving for sense objects", which is usually translated as "sexual desire". It is to note that unlike Judeo-Christian myth where sexual desires are considered to be the "fall of man", Hindu dharma doesn't associate sex with shame, instead, a joyous aspect of human existence. This is where Kamasutra comes into play, which is not just about positions, but more about the philosophy of love and the questions of love.

Shringara – Rapturous Intimacy

This is where the Indian sages focused on the emotional content and developed a rich context to understand the moods and emotions.

This is where 'Shringara' was born, which is romance. The erotic attraction from just sensory cravings is stirred up when they share secrets with each other, make affectionate names of another, play games, exchange inventive gifts. This sort of imaginative play of love is symbolized by the relationship between Radha and Krishna, where they were involved in adventure, music, dance, poetry, and many other romantic gestures.

However, the ancient sage understood that just finding our "soulmates" would not be able to solve all our problems. However, it does provide us with the foretaste of something good, something much greater.

Maitri – Generous Compassion

The dating sites and apps are screaming: "Stop waiting for love; it is within your power right now to make it happen." But can just a swipe of a finger or a click of a button really help you on your spiritual journey? Maybe, maybe not.

Because there is little love in little things we can do. This could just be a simple smile to a stranger or food to hungry. Mahatma Gandhi says "The simplest acts of kindness are by far powerful than a thousand heads bowing in prayer." Compassion just like the natural love we feel towards children and pets. The "Matru-Prema", which in Sanskrit means motherly love, and it is the kind of love that is most giving and least selfish. Maitri is the motherly love, not just for your partner or children, but for all living beings.

Compassion for everyone might be hard for all. Thus, there are practices in Hindu and Buddhist dharma that can help you develop this ability. And when you are more compassionate with everyone, you become even more compassionate with your partner.

Bhakti – Impersonal Devotion

Compassion is just a start, and not a final rubber stamp. The ancient sages went beyond interpersonal love and envisioned a form of love that expanded to the whole of creation. This path to finding this love is known as "Bhakti Yoga", where you are not only cultivating the love for others but also to the love of God. The love of God doesn't need to be the God that we use in the usual self. It can be the highest form of kindness, truth, or justice.

Atma Prema – Unconditional Self-Love

There is a cycle to everything. The love that started by directing towards others is then again directed towards the inside – the self. Atma Prema could be translated as self-love. Indian mystical poet Kabir says, "The river that flows in you also flows in me." This means that what we see in others, we see others in ourselves.

Once we achieve self-love, we recognize that we are all one, in that we recognize that our genetic heritage and upbringing have little value in the distinction of our self. The true self is all.

RELATIONSHIP CONCEPT AND WHICH TYPE OF PARTNER WE CHOOSE

According to me relationship is not just about 2 persons. That one is not a relationship which currently people thinks that (boyfriend & Girlfriend). That is not relationship that's just a stuff fulfillment of desires before marriage like money, lust, etc. According to bhagavad Gita that also sin. So please be careful about it that is not relationship. The concept of "relationships and family" is broad and varies from person to person. ... In our model, personal relationships refer to close connections between people, formed by emotional bonds and interactions. These bonds often grow from and are strengthened by mutual experiences.

A relationship is any kind of association or connection between people, whether intimate, platonic, positive, or negative. Typically when people talk about "being in a relationship," the term is referencing a specific type of romantic relationship involving both emotional and physical intimacy, some level of ongoing commitment, and monogamy (i.e., romantic and sexual exclusivity, wherein members don't have this type of relationship with anyone else). That said, romantic relationships can take many different forms, from marriage to casual dating to ethical nonmonogamy.

Basic types of relationships:

- **Familial relationships, aka family members or relatives**
- **Friendships**
- **Acquaintances**

- Sexual relationships (That one is just sin or stuff. Which makes your soul impure and week)
- Work or professional relationships
- Teacher/student relationships
- Community or group relationships
- Place-based relationships, such as neighbors, roommates, and landlord/tenant relationships
- Enemies or rivals
- Relationship to self

Which type of partner we need to accept in our life

In today's time you don't have to go to the Himalayas to find spiritual awakening. Your relationship can be as deep a well of spiritual growth if you find someone who helps you in spiritual path. as you'd find on any meditation retreat. A spiritual relationship is one of the most powerful tool for transforming you into your best self. Choose someone who is spiritual it's for both (man or woman) if you really want to live better life and really wants to evolve then you try to find spiritual person. But be careful in today's time there are lot's of fake scammers who acting like spiritual but they really don't know about spirituality.

I will recommend spiritual person because they have some qualities I also seen in mine too that's why I recommend spiritual person only.

- They live balanced life beyond happiness and sorrow. Spritual person's mind also pure. They never thought any negative things or any stuff's that makes a person crazy.

- They have Steady mind that's help them to go beyond happiness and sorrow. They never drink, smoke, eating non-veg, etc.

- Anger free they are always. They really don't care about what other's thinks about them.

- Spiritual person will accepts simple lifestyle always. They really don't believe in showoff of material things.

- Because of that steady mind they really free from lust and other material attraction.

- By nature they really kind and polite. they never talk loud and never use dirty words.

- Dressing sence also very simple of spiritual person and they always respect other. They always offers there all things to lord (God).

- Spiritual person always free from all types of sins like lust, eating non-veg, using dirty words which is on trend in today's time people thinks that make's them cool, etc.

Every relationship has its own unique set of challenges, I know it's really hard to find someone like that. But need to try maybe maybe you will find. It's for those who really wants to evolve themself and want better life in future.

How we judge person is spiritual.

So when you meet someone check somethings like:

- first check dressing sence and hair style too Spiritual person really looks simple. they far away from showoff and other material stuff's.

- Check his talking style Spiritual person really talk's respectful and politly. I understand in today's time it's really hard to believe anyone totally. In beginning all talks politely and with respect so maybe you can't able to understand him/her properly.

- Check and try to understamd his/her Thoughts when you talking with him/her.

- Before meeting someone try to verify that person's background. That's also important.

- Try to check his behaviour. Spiritual person always behave's simply. Spiritual person really very decent by nature.

- A great partner is someone who thinks not only of you but of others as well. Pay attention to how they treat the people in their lives and nearby, from their peers to their parents to the people they encounter in everyday life.

- Does not make judgmental comments about others and also check if him/her attracting because of looks & lust. Try to avoid him/her. because if you accept him/her in future you will get pain defenitly. No doubt about it.

- Gives genuine compliments about anything.

- Is patient with service people, including waiters and cashiers. Treat them with respect.

- Shows their appreciation for others.

- Thinks about how others are feeling.

Try to live with your partner like Ardhanarishvara-The combined form of Shiva and Shakti

Ardhanarishvara is an androgynous form of Lord Shiva, the composite of Shiva and Parvati, the half-male and half-female fusion from the center. The right half is that of Shiva, and the left half expounds Parvati.

Symbolizing the all-pervasive, all-enduring nature of Lord Shiva, Ardhanarishwar portrays a perfect balance of male and female as equal, as "Purusha" and "Prakriti," the feminine and masculine energies of the cosmos as well as Shiva, the male principle of God, as the essence of Shakti, the Sacred Feminine.

Ardhanarishvara is also named Ardhanarisha, Ardhayuvateeshwara, Ardhagureeshwara, Gaureeshwara, Naranari, Parangada, and Ammiappan. According to Shaiva Siddhanta, Ardhanarishwar is one of the 64 manifestations of the absolute Parashiva.

The origin of Ardhanarishvara

There are different instances about the creation of Ardhanarishvara according to scriptures and different Puranas. Among all those two are the most discussed ones.

There was once a Rishi called Bhringi, a devotee of Lord Shiva. He used to believe himself to be the greatest acolyte of Lord Shiva only to the bit that he refused to worship Shiva along with Parvati. He had solely dedicated himself to Lord Shiva but would not worship his consort Parvati.

One day, Rishi Bhringi reached Mount Kailash, the adobe of Lord Shiva, to circumambulate Shiva, but he repudiated to circumambulate Parvati despite her being alongside Shiva. Goddess Parvati then urged Shiva to unite themselves together. That's how Ardhanarishvara was created, one half of Shiva and the other half of Parvati through the central axis.

Rishi took the form of Beetle and circumambulated Shiva only, which enraged Parvati. Parvati then cursed Bhringi to lose all the blood and muscles believed to have come from mother in Hindu embryology. Bhringi was all but the only skeleton now, which is believed to have come from his father, making him realize the significance of Prakriti and Purusha. He pleaded forgiveness from Parvati and was given the third leg as a reward for pleading to sustain his body.

According to Shiva Purana, Lord Brahma, the creator of the universe, was disappointed with his creation as the world was not moving at the pace. It was constant to the number of beings he had created. There was no way out for him rather than calling Shiva for speeding the development in the world. Brahma asked Shiva for help, and Shiva took this Ardhanarishwar form to make him understand generation through copulation. Afterward, Ardhanarishwar split into Purusha and Prakriti, thus continuing the creation, suggesting that Shiva is nothing without Shakti, and creation, as well as the continuation of life, is impossible without both of them.

Symbolism of Ardhanarishvara

The apologue of Ardhanarishvara has an intense meaning symbolizing the quintessential balance between the male and female energies in the universe. The forces are inseparable and complementary to each other, suggesting that they must work together to maintain the equilibrium.

The unity of Purusha and Prakriti are basically opposite in nature. Still, they match each other in the sense that Purusha is the passive force of the universe. At the same time, Prakriti is active, Purusha represents potential energy, and Prakriti is kinetic energy, and Purusha is infinite. Prakriti brings that infinity to be finite, thus, embracing one another to generate and sustain the universe.

As mentioned earlier, Ardhanarishvara is half-male and half-female; however, this does not mean Shiva and Shakti are incomplete. Ardhanarishwar signifies that "totality lies beyond duality"; they are complete and fully developed man and woman. They are supreme and equal, which is why they both are the essence of creation. Shiva on the left also signifies the spiritual sphere and Shakti the material sphere depicting that both must coexist to bring ecstasy to life.

Many cultures also believe that Ardhanarishwar marks limitless growth and fertility as Parvati with Shiva is associated with the profuse reproductive ability of mother nature. Generally, the Shakti half is on the left side, denoting relative inferiority and feminine characteristics like creativity, intuition, etc. The right side is of Shiva, half denoting comparative superiority and masculine characteristics like strength, logic, systematic thoughts, etc.

There are some characteristics of spiritual relationships that any seeker should need to cultivate.

You both practice compassion in moments you might be tempted to judge.

In those moments of conflict, or in struggles you or your partner face individually, you don't criticise or affirm each other's negative self-perceptions. Instead, you listen with understanding and acceptance. You see your partner's wounds as something with the potential to transform them through healing, and they see yours the same way. You have faith in each other's ability to surmount these challenges and encourage each other to do so.

Doing meditation together and try forgive each other.

That will helps you both to live peacefully When conflicts arise, you try practicing meditation. That will makes both of you minds stead. You never use any harsh words and hurt feelings to each other. That will make both of you mind pure and you will try to forgive him/her if one of you did something wrong.

Your only expectation for each other is that you each try to be your best self every day.

There's really only one reason relationships fail: It's that the expectation of one or both parties did not align with the reality they experienced. It doesn't matter what the expectation was—better communication, a shorter courtship period, that your partner wouldn't change—expectations are the doom of a relationship if you let them define it.

You are fully present in the relationship.

In a spiritual relationship, partners show up completely—emotionally, spiritually, mentally. You do not hide from each other and you feel comfortable being vulnerable and truthful to each other. (But you do it even if you don't.) You invest in self-awareness practices like meditation, yoga, and journaling so you are constantly increasing the awareness of what you want and need, and what you can do to be a better partner. Then you practice conscious communication to make sure your partner hears those needs. You both speak and listen with the intent to understand each other.

Try to avoid Sexual activites that will make your soul weak and impure.

NIRVANA

What is Nirvana?

It is a state of Ananda, eternal bliss, and joy. It is overcoming the ignorance that we live in and being enlightened with the Truth. It is a state of self-realization, a state that liberates us from misery and sorrow while we are alive. Not only that, it also liberates us from the cycle of death and rebirth. Unlike Moksha which has no author, Nirvana is the brainchild of Prince Siddhartha Gautama who went on to be known as the Buddha or the awakened one. Tired of the inhumane practices during his time, he renounced his kingdom, all the wealth and luxury and went in pursuit of the truth. He saw so much suffering, so much cruelty that his heart was torn into pieces. Being full of compassion, he went from sage to sage, monk to monk, from one ashram to another monastery searching for the truth.

To him, nothing else mattered. Being a Hindu Kshatriya, the royal class, he was tired of the influence of the Brahmins, the priestly class who had manipulated theology to suppress the lower castes.
He questioned every ritual and superstition until he finally awakened to the truth. Let us try to understand how Nirvana came about. Around 1000 BC, the Hindu Faith also known as Sanatana Dharma, that prescribed Moksha to be the Ultimate Goal of life started getting diluted into Hinduism. Which I already metioned on Earlier, God was considered to be a power that was omnipresent, omnipotent, omniscient but by now, Hinduism hadbecome a religion of rituals and superstitions. It became a religion of innumerable Gods with name and form.

The priestly caste or the Brahmins became very powerful, even more powerful than the kings themselves. To keep their power supreme, they created several rituals and superstitions that started making Hinduism unpopular. These included the caste system, which made life for the poor people miserable.

Nirvana became a solution to all the ills that developed in the Hindu belief system. When the Hindu Prince of Kapilvastu, Siddhartha Gautama who later became the Buddha, grew up to be a young man, he questioned the supremacy of the Brahmins just as he wondered why so many illogical rituals and superstitions were being followed. While he grew up in a palace and was protected from all the signs of suffering, it is said when he was born, he was destined to become a world conqueror, either he would be king of kings or he would be the most realized spiritual leader who would win over humanity with his love and compassion. This was a matter of great concern to the king and queen.

Since a noble sage prophesied that the prince Siddhartha Gautama could become the biggest saint humanity has ever seen, his father, the king was disturbed and made all-out efforts for the prince to live in the lap of luxury so that his mind never entered into the realm of spirituality, but as destiny would have it, what the sages had proclaimed became true.

The king did everything he could to keep his son away from the thoughts of suffering. He even made a new city where the sick and the old were shifted so his son would never see the dawn of suffering in life. But eventually, the prince came face to face with suffering and death. What were the various realizations of the prince before he was enlightened to be the Buddha? The Prince had a very kind and compassionate heart. Although he was a very happy person, he felt saddened by the suffering that he saw. He is said to have observed 'four signs' which led him to his quest. First, he saw an old man, which made him realize that life on earth is not permanent. We will not remain young and healthy forever. None of us can escape old age. He realized that the body will become weak and this will ultimately lead to suffering.

He then saw a sick person and was distraught to see the suffering in this world due to disease. As a prince, he had been sheltered from seeing such pain. He realized that most of us will suffer as our bodies will face disease and decay. He went on to see a corpse that made him realize that life is not a continuous process. We all have to die one day and with death, we will lose all our pleasures and possessions that we believe belong to us.

While he saw these three signs of suffering, he also saw a monk sitting in meditation peacefully beneath a tree. The peace and bliss on the face of the monk inspired him to realize that it was possible to live with joy and tranquillity. Unfortunately, he had not been exposed to these realities of life.

The prince did not become the Buddha overnight. His quest led him from being a seeker to a master. The prince Siddharth did unusual things, quite unlike a king. He refused to go to war as it would lead to a loss of so many lives. He was compassionate and for him winning a kingdom was not as important as saving lives was. The turning point in his life was when he saw a war, in which hundreds of people lay dead on the battlefield. He was shocked to see the suffering and bloodshed and wondered why. He realized the futility of war and the need for peace. Seeing the horror, the prince said, "I don't want this wealth, this kingdom, I don't want war, I don't want power. I want peace, I want Liberation from this misery, this pain."

Once, the prince was asked by saints to kill an animal as a sacrifice as per the scriptures but he refused to do it. Not just this, he refused to do anything that was inhuman and lacked compassion. He was against rituals, superstitions and gave utmost importance to kindness and fairness as he looked at each issue practically and with human eyes.

In his quest for the Truth, Siddhartha went on to become a Sanyasi, a renunciate and went deep into meditation. As the future king of Kapilvastu, Prince Siddhartha Gautama was not attracted by the kingdom and the wealth that he possessed. Even though he had a beautiful wife and a lovely son, he left his family, mother and father included, one night to go into the forest in search of the truth. The prince realized that to be peaceful, one must live with Ahimsa or non-violence, a philosophy that believed that we must not cause any injury to anybody in thought, word or deed. He advocated that we must forgive unconditionally. Every human being makes mistakes in life but the moment we realize our wrongdoings, we are forgiven by the Divine.

After that, Siddhartha Gautama became a Tapasvi, living a life of sacrifice and deprivation. He even gave up eating fruits and drinking water to deprive the senses of the body, he gave up all desires, for the purpose of achieving Liberation. In his quest for the Truth, Siddhartha Gautama happened to meet some of the followers of Mahavira, a sect that followed the Jain Tirthankaras. To them, life was all about Tapasya, letting the body suffer physical pain and living with deep austerities. He himself went into such a practice till it is said that he nearly died as he deprived his body of food and water. A young maiden exclaimed, "You have become like a stick where your front and your back have become one. What is the use of such a life?" She asked the prince who was in search of the truth, "How would such deep Tapasya and austerity help in realization and Liberation? If you make the string of the sitar very loose, you cannot extract music from it, just as tightening too much will snap it." The Buddha realized that such extreme austerities would not help us to achieve the goal of life. He realized that one cannot realize the Truth by making the body suffer. It is through the body that one achieves Liberation. He then coined a new way known as the 'Middle Path' philosophy for people to live a balanced life.

The way to realization is through self-effort. Realization is intuitive. A Guru can guide but cannot give Mukti. Siddharth was called the Buddha because he realized the Truth by his own experience, going within. He told his fellow monks, " There is no need of doing sacri ces and Tapasya. We must not torture the body. This is Avidya or ignorance. We must live with the realization of the truth. We need to realize ourselves." "Our biggest problem is our ego. Due to our ignorance, our ego creates worry, fear, and misery by believing in illusions following rituals and superstitions which make no sense. Ignorance creates arrogance, it makes us angry, we worry, live with fear and anxiety all due to our ignorance."

The Buddha called his realizations the Four Noble Truths of Life. What were the four noble truths?

1. Dukkha – the world is full of suffering.

2. Samudaya- the origin of suffering is desire.

3. Nirodha- if we give up desire, we can escape suffering.

4. Magga- there is a path to follow to renounce desires, which ultimately came to be known as the Eight-fold Path.

Let us understand what the four noble truths are. The first was the existence of suffering or Dukkha. The sights that the Buddha saw as a prince shocked him, just as they made him realize that one cannot escape from suffering. As long as the universe exists, misery will follow us like our shadow.

When he went on a quest to understand why we humans suffer, he realized the second Truth of the origin of suffering Samudāya. He advocated that suffering was mostly caused by desire or craving or Trishna. The cause of this misery is attachment or craving and desire for somebody or something. It is important to activate our intellect and realize the truth, not to react to any joy or sorrow with too much passion. If we follow the right path, we will be peaceful, blissful and we will be liberated from this world of misery and from suffering. The Buddha taught that the way to extinguish desire, which causes suffering, is to liberate oneself from attachment. This is the third Noble Truth - the possibility of liberation, the cessation of suffering or Nirodha. Was there a way to overcome this suffering? Yes. One could live a life of bliss or Nirvana. The nal Noble Truth is the Buddha's prescription for the end of suffering, the path to the cessation of suffering or Magga. However, he taught us that we cannot achieve Nirvana unless we follow a particular path, which he later called the Eightfold Path. What was this Eight-fold Path?

- **Right Understanding: The Buddha taught us that we must not accept anything without proper discrimination.**
- **Right Intention: The Buddha advocated that our intentions should be pure, not cunning or manipulative.**
- **Right Speech: Whatever we speak must be the truth and must bring about positive consequences. We must not hurt anybody with our words.**
- **Right Action: We must not do anything that harms others. Our actions must only be for the good of humanity.**
- **Right Livelihood: We should make a living through ethical means, living with values.**
- **Right Effort: We must control thoughts and feelings from leading us to evil actions.**
- **Right Mindfulness: Living with awareness and consciousness of the truth and nothing else.**
- **Right Concentration: Training the mind to think, introspect and contemplate the right and pure thoughts.**

The Buddha described the Eight-fold Path as a means to Nirvana, like a raft that is needed for crossing a river. It was by following this path, said the Buddha, that we can cross the river of Samsara. After becoming the Buddha and being blessed to realize the Truth, being the 'Awakened One,' the Buddha dedicated his life to helping people live a life of peace because all around him, he saw people suffering and living a life of ignorance with stress, worry, and anxiety. He created 'the Sangha' and welcomed people to join it by shaving off their hair, wearing orange robes and following the rules of the Sangha. While he made certain rules to join 'the Sangha', he did not expect everybody who followed his teachings to do so. He went from city to city, kingdom to kingdom, teaching people to live a life of peace and nonviolence. Whether it was the richest king, a religious priest, a rich businessman or a low caste scavenger, to him all were equal. In those days, women were not allowed to evolve spiritually or take important positions, but the Buddha changed even this belief because he believed in the concept of equality. He welcomed everybody to join his 'Sangha' including women.

In the times when the Buddha lived, there was so much disparity in the society because of the caste system. Society was divided into four castes – the Brahmins or priestly people, the Kshatriyas also called Rajanyas who were the rulers, administrators, and warriors, the Vaishyas or artisans, merchants, tradesmen, and farmers, and Shudras or labouring classes, the poorest of the castes. The poor and the weak in society were tormented and tortured. He questioned why this should be so because all humans are equal. He opposed the concept of 'untouchables' that existed in the Hindu society then, just as he questioned the supremacy of the kings and the Brahmins. According to the Buddha, to look down upon another human being based on caste and creed is a shame. Although the scriptures and the society in those days strongly followed the caste system, the Buddha openly opposed it and gave importance to the equality of all humanity.

Buddha's teachings and thoughts destroyed this caste system with the logic that all human beings are the same, our tears are the same just as our blood is. The skies and the clouds, the mountains and the streams, do not differentiate in the way they treat humanity. If nature doesn't differentiate, then why should a man do so? Buddha said all humanity is equal. We must not divide people into high caste and low caste.

The Buddha, thus, accepted untouchables as a part of his following. Although he knew there would be great opposition from the priests and the kings, he boldly renounced all divisions of society as per caste. This was one major change in society at that time. As the prince, the Buddha was very friendly with king Bimbisar, king of Magadha, and the king continued to be his friend even after Sidharth Gautama became Buddha. King Bimbisar changed his entire style of ruling the kingdom and implemented the teachings of the Buddha setting aside the advocacies of the priests, even at the cost of antagonizing them. One can follow the Buddha's path by living with the five virtues that the Buddha advocated. All the religions of the world are based on the fundamental principles of good conduct and prohibit their followers to indulge in the misconduct and misbehaviour that may harm society at large. So, the Panchshila of Buddha comprises the basic teachings of conduct which are as under:

- **We should never kill anybody, not even an animal.**
- **We should never steal.**
- **We should not live with adultery.**
- **We should not lie in any way.**
- **We should not consume any kind of intoxicants.**

When the Buddha was asked how one should live, he said that we should follow the five rules of life. Buddha also said that not everyone needs to become a Bhikshu or monk to realize the Truth. Somebody asked Buddha, "Why do you beg. Why can't you farm your own food? He replied, " I'm also farming my soul and with the grains, I will feed Souls." It's not that Bhikshus do nothing. They do far more important work than just farming or other work. They save people from misery. Helping people to be liberated is more important. The Buddha was very practical and prescribed a lifestyle that would make one blissful and peaceful. The Buddha lived a life that set an example for his followers. He chose humility and non-violence over victory and power. He transcended the body, mind, and ego with his compassion and love. Through his sacrifice and renunciation, he inspired people to embark on a journey of ultimate peace and joy. He showed sympathy even to his biggest enemy because he realized the true meaning of life. The Buddha explained how anger burns our peace eace and happiness. We spend time in unnecessary arguments and then, create poison within our own hearts and become miserable. When we get angry, we throw burning coal at others to destroy them, but before that, the burning coal burns our own hands.

The Buddha said: "Hatred can not be defeated by hatred, it can only be overcome by means of compassion. It is the rm and unchangeable Law of Nature."

He gave a lot of importance to contemplation and realization of the Truth. In his life, there were several instances where he transformed people, sinners, and criminals by his spiritual wisdom. The Buddha was very clear that all our suffering was due to ignorance. We experience suffering from birth to death. To be happy, we desire, we crave, we have expectations from others, and ultimately, we are disappointed. This distances us from the bliss that is within. He advocated that we must nd peace and happiness by going within. Why do we chase money and pleasure in the journey of life that is so short? Instead, we can live with love, compassion, kindness and create a blissful heaven wherever we are. We create our own fears, worries, and miseries by our own mind. One who transcends their mind, said the Buddha, has conquered the whole world. It is most unfortunate, he said, that we do not understand what the true and Ultimate Goal of life is. He termed that state as 'Nirvana'. The only way to achieve Nirvana is through the realization of the Truth, contemplating spiritual wisdom that will liberate us, and living with soulful love for one and all. We think we are everything when in reality, we are nothing. This is nothing but ignorance. What does ignorance do? It makes us live a life of illusion. It creates worry, stress, and anxiety which is the cause of all our misery. What is ego? The ego gives birth to wild sensual desires of the body, just as it permits the mind to create miserable thoughts. When we realize the Truth of the cosmic illusion, we experience a state of everlasting peace. We live with compassion and love, despite all circumstances and evil-doing of those around us. Nothing affects a realized soul, explained the Buddha. We must activate our intellect to overcome miseries - that is the way to escape from fear and worry. We must live in the present moment. The way to freedom from misery is to live a life of Divine acceptance and surrender. We all want to be happy, just as we want to be liberated from misery. But we choose actions that take us in the opposite direction. How can we achieve our goal? We become slaves, prisoners to our own desires and expectations. Although we may look peaceful on the outside, we experience violent turbulence in the inside, with our own thoughts destroying our peace. What did the Buddha have to say about death? He accepted it as a reality of life. It should be accepted without expression of sorrow. All those who are born are certain to die. Nobody can escape death. Then why should we become unhappy when somebody dies? This is the law of nature. Our life is like the moon – it appears as a new moon and then becomes a full moon, after which there is complete darkness.

Then the moon reappears, so is our life. Death is not the end. Only the body dies. We will be reborn as per our Karma until we are liberated. Buddha made people realize that death is certain. Let us learn to accept the reality of death. Because we become attached to people and things not realizing that nothing is permanent, we live in misery. How does one escape from this misery and achieve Liberation? The only way is to live by our intellect and not by our mind. We must learn to discriminate the illusion from the Truth, and thus live a life of Realization. As long as we don't accumulate the wisdom that makes us realize the Truth, we will continue to suffer. Suffering is a reality, we cannot escape from it. But the truth is that suffering is caused by desire and craving. If we eliminate the cause of suffering, then how can suffering continue? The Buddha taught that prayer by itself can achieve nothing. Of course, prayers are magical, but they must be accompanied by Divine action, meditation, and education that leads to realization. Just following certain rituals and superstitions can get us nowhere. It is possible to achieve Ananda or bliss and be liberated or achieve Nirvana. However, this needs our effort. It needs discrimination and a quest for Realization. We must light our own lamp to illuminate the darkness within, just as the lamp will create a path to move forward. We must be conscious of the power that is within us. We don't have to search outside. Everything is within. We must be steady in our discipline and thought, and this must be our persistent effort until we achieve Nirvana. Buddha advocated that the world is a Cosmic Illusion or Maya. It robs us of the present. We don't live in the past, nor in the future. We should be conscious of the present moment following the 'middle path.' We should not torture the body, nor give it scope to crave and desire. The moment we realize the illusion of Maya, we end our worry, fear, and anxiety. We live peacefully and blissfully. As we realize the Truth, we don't feel hatred against anybody. Even a person who is cruel to us is responded with compassion and kindness. That is true realization. Realization makes us live with love and Divine Acceptance. We get calm and peaceful with a steady intellect. It is our ignorance that has separated us from each other and has created a division. Thus, we desire, we crave and we are lost in the ignorance, in the illusion. When we realize the Truth, ignorance is overcome.

The Buddha said: "The biggest enemy of a man is his own mind which makes him do inappropriate tasks. What we think, we become. A good thought brings in good results while a bad thought brings in bad results. This is the Law of Nature."

The Buddha said, "We are living in utter ignorance. We perform rituals and follow superstitions, we sacrifice the lives of innocent animals but ourselves continue to live with sin, how can we achieve Liberation?" He said that we are the cause of our own Liberation or our own misery. It's for us to realize the truth by living a life that leads to Nirvana. The Buddha stated: Ignorance is the root cause of all suffering. Ignorance cannot be overcome by worshipping, fasting or offering. It can only be eradicated through meditation, which will result in wisdom and realization of the Truth. The Buddha said, "We are spoiling the present moment with what happened in the past and worrying about the future losing the present moment." To be liberated one must not overreact to misery or to joy. Accept both equally. People think the Buddha said, "There is no God." But he never said that there is no God nor did he say there is God. Buddha said that prayer without efforts will not work. If you want to cross a river and you sit down on the bank of the river and pray, would you be able to cross the river? Of course not. If the water is shallow, you can walk. If the water is deep, you can take a boat or even swim. The Buddha said that if one doesn't want to walk across, doesn't want to swim, or doesn't want to use the boat, will the other side of the riverbank come to him? Similarly, if we don't use our intellect to overcome ignorance, how will we achieve a life of peace and bliss? How much ever we pray, do rituals, follow superstitions, we cannot achieve the objective that we seek without using our intellect. This is not dificult to understand but we don't understand it because of our ignorance. Prayer is good but without knowledge, one cannot realize the truth and be liberated. He said that we don't have to have blind faith. We should use our intellect to discriminate. Somebody asked Buddha, "What did you get out of Meditation?" He said, "Nothing! Just that I lost worry, fear, anger, sorrow, insecurity, the anxiety of old age, and death." Human beings get attached to their possessions. They are never satisfied. They crave for more and are fearful of losing what they have. But if we cut away all attachments then we are free.

The Buddha always said that one should become one's own lamp and light one's own light. Which means, one should try to seek the reality oneself and realize the truth by making self-efforts. Instead of accepting anything blindly, one should try to question everything and then accept what is right. The Buddha was a very practical person. He told people not to believe anything, not even him. He said that we should question everything with our intellect and use our logic to realize the Truth. While he refused to follow any ritual or superstition, he encouraged people to question their religion and idol worship. His objective was simple, to be liberated from misery and to live a life of eternal peace and joy, attain a state he called Nirvana, our Ultimate Goal. According to him, we must live a life that ultimately liberates us, a life of salvation and enlightenment. What was the key difference between Moksha and Nirvana? The Buddha might have called it Nirvana but it was no different from Moksha. The destination was the same, the means to reach it was different. During the Buddha's time, Hindu Faith had lost its original principles and declined to a religion of Gods and Goddesses, of rituals and superstitions, of caste systems and division of society. The Buddha opposed all this and prescribed a simple path to the same goal of Liberation which he called Nirvana, the Ultimate Goal of life. It was all about living in bliss, peace, and joy, liberated from misery and sorrow. This was no different from Moksha.

The Buddha's logic was very simple: Why should we worry about things which are beyond our comprehension? Why don't we focus on our Ultimate Goal of life, peace and bliss? We want to create Gods and Goddesses to pray and instead of understanding the Creator who created us, we spend our life in rituals and superstitions. Instead of overcoming our ignorance, we are creating more reasons to become more ignorant by complicating our life. What do we human beings seek? Our Ultimate Goal is happiness and peace. Why not achieve that? Why are we submitting to the world of illusions and becoming miserable? Why are we complicating our life with desires and expectations and becoming miserable? The Buddha was against all this. He had a simple solution. He called it Nirvana - ultimate joy and peace, a life that is liberated from misery and sorrow.

POWERS BEYOND HUMAN BODY

If someone really want some super powers then its possible. In ancient time there are many sages. That time they had a lot's of power (Sidhi or Nidhi). If anyone will work on Spiritual path properly then the person can able to attain some spiritual powers beyond human body but that's not one day work maybe it will take 1 year or 10 years time. It depends on person's abilites and character also matters in it.

Before work on Spirituality you need to try these things:

- Before start working on spiritual path you need clear your mind. Crying also very useful that will make our mind light. Actually When we cry the level of endorphins, Enkafalin, prolactin, Lucaine would be low so that will help in reducing stress. You need to clear all types of stuffs in your mind. Make your mind pure.

- Bath before that and then worship god by mantra and use Sandalwood on your forehead. (I recommend Sandalwood because putting Chandan (Sandalwood) have there own benefits like Relieves from insomnia and stress, also effective in home remedy too.

- After these things then start working on spiritual path. Person will definitely feel some changes on body and mind. First you will feel stead mind then when you go deeply you will feel it inside your body but that will work if you do that properly and don't expect it on first day. It will take time. Also try to avoid sexual activities that will make your soul weak and impure.

What exactly do we mean by these Siddhis and Nidhis?

Ancient sages had developed these (Spiritual) Adhyatmic powers (Powers of the soul) with two main objectives in mind,

the first objective was to make human interaction on this planet to not be of the parasitic nature that it is, you know the way the human existence on this planet today is of a parasitic nature, these sciences aimed at changing that.

the second reason for which these powers were developed were to obviously gain exceptional powers which a human would not ordinarily gain.

The powers that can be obtained by these sciences are called Siddhis and Nidhis

The eight Siddhis:

- **Anima**- Reducing one's body to the size of an atom
- **Mahima**-Increasing the size of one's body into an infinitely large size
- **Garima**- Which means being infinitely heavy in weight
- **Laghima**- means becoming almost weightless
- **Prapti**-means having unrestricted access to all places, this is something that Lord Narada has, the ability to go anywhere in the Universe
- **Prakamya**- means realizing whatever one desires
- **Istva**- means possessing absolute lordship
- **Vastva**-means the power to subjugate all.

Now, Five siddhis of yoga and meditation as given in the Bhagavata Purana are, the five siddhis of yoga and meditation are:

(1) **trikālajñatvam**: knowing the past, present and future, which is a Trikaal Darshi Rishi or sage.

(2) **advandvam**: tolerance of heat, cold and other dualities, so you see making human body independent of the climatic conditions around it. This is a part of what was mentioned earlier, a part of the Ancient Sciences which made human dependence on the planet to not be of parasitic nature. So the first thing you see is, they developed a science so that the climatic conditions around the human do not affect him

(3) **para citta ādi abhijñatā**: knowing the minds of others and so on

(4) **agni arka ambu viṣa ādīnām pratiṣṭambhaḥ**: checking the influence of fire, sun, water, poison, and so on

(5) **aparājayah**: remaining unconquered by others

Secondary Siddhis mentioned in the Bhagavata Purana are, Ten secondary siddhis given in the Bhagavata Purana, Lord Krishna describes the ten secondary siddhis:

(1)**anūrmimattvam**: Being undisturbed by hunger, thirst, and other bodily appetites. So you see this is the second example of making human body independent of the resources of the planet, one is where the human body is undisturbed by the climatic conditions of the planet and the second one is where the human being is undisturbed by hunger, thirst and other bodily appetites, ok, so you can imagine that a person who does not need food, does not need water, who is unaffected by the climatic conditions around him, you can imagine that his existence on this planet would not be that dependent on the resources of the planet as a normal human being. Now next one is,

(2) **dūraśravaṇa**: Hearing things far away, now this Siddhi of Dursravana is a Siddhi that Mr Nikola Tesla seems to have had, it is mentioned in his autobiography "My Inventions" that he could hear things from 20 feet away and since he was a karma yogi he had developed these powers.

(3) **dūradarśanam**: Seeing things from far away,

(4) **manojavah**: Moving the body wherever thought goes. Now please pay attention Manojavaha is the power to move the body wherever the thought goes, now this power, this moving the body wherever the thought goes it menas travelling faster than the speed of light, you know, one can be sitting here and think of Mars and be there in a matter of half a second, you know, less than half a second, you know, so what we are talking about here is a speed that is faster than the speed of light which is the speed of thought.
(teleportation/astral projection)

(5) **kāmarūpam**: Assuming any form desired

(6) **parakāya praveśanam**: Entering the bodies of others

(7) **svachanda mrtyuh**: Dying when one desires,

HUMAN BODY AND UNIVERSE ELEMENTS

The human body is made up of elements in the following approximate proportions (by weight): 65% oxygen, 18% carbon, 10% hydrogen, 3% nitrogen, 2% calcium, 1% phosphorus, and 1% other elements such as potassium, sodium, iron, zinc, etc. By the number of atoms, however, the proportions are: 63% hydrogen, 24% oxygen and 12% carbon, with only tiny traces of the others.

The earth's crust is made up (by weight) of: 46% oxygen, 27% silicon, 8% aluminum, 5% iron, 4% calcium, 2% sodium, 2% potassium and 2% magnesium, plus traces of the other 84 naturally occurring elements. The air we breathe contains roughly (by volume): 78% nitrogen, 21% oxygen, 1% argon, 0.038% carbon dioxide, and trace amounts of other gases.

The Sun is composed of: 75% hydrogen, 24% helium and 1% oxygen, with tiny traces of carbon, neon and iron. In fact, hydrogen and helium are estimated to make up roughly 74% and 24% respectively of all the matter in the universe as a whole, along with tiny amounts of oxygen (1.07%), carbon (0.46%), neon (0.13%), iron (0.109), nitrogen (0.1%), silicon (0.065%), magnesium (0.058%) and sulphur (0.044%).

<u>The universe exists within us, and we exist within the universe.</u>

For thousands of years mankind has been allured by fascinating, vedic, greek or anyother concept. Each and every technological invention has been an attempt to reconcile our inner desires with our outward experiences.

Technology has always been a product of ingenuity and curiosity. Although a great tool for exploring the unknown, it still has a long way to go. The human body exists within a universe whose fundamental laws are still largely a mystery, a subject ripe for research. With its laws, its order, its chaos, its mysteries, the body is a mirror of the universe.

Throughout history, doctors have always used the technologies at their disposal to read this universe – without ever forgetting the most important tool of all, one beyond the reach of any artificial intelligence even today: intuition.

Esaote's purpose has always been to create instruments that help doctors read our physical universe more clearly – and, like any good doctor, it has sought to better itself over the thirty years or so of its existence.

Research and development, without limits

The importance Esaote places on discovery is evident from its various collaborations with scientific-medical and technological research centres, as well as space centres. Esaote is always present where it is needed, accompanying mankind into new, uncharted environments, following the rhythm of human curiosity. As mankind forges ahead, Esaote stands at the forefront, keeping up with the ever-increasing pace of human progress. Over the past year, Esaote's ultrasound range has been completely updated, with mobility, lightness, precision and sustainability requirements in mind. Existing technologies are no longer enough to cope with the diversity of diagnostic specialisations and conditions of use that exist today. We must go further, all while taking concrete scientific knowledge as the solid foundation for our development.

www.ingramcontent.com/pod-product-compliance
Lightning Source LLC
Chambersburg PA
CBHW070810220526
45466CB00002B/622